For John and my parents

The Sierra Club, founded in 1892 by John Muir, has devoted itself to the study and protection of the earth's
scenic and ecological resources — mountains, wetlands, woodlands, wild shores and rivers, deserts and plains.
The publishing program of the Sierra Club offers books to the public as a nonprofit educational service in
the hope that they may enlarge the public's understanding of the Club's basic concerns. The point of view
expressed in each book, however, does not necessarily represent that of the Club. The Sierra Club has some
sixty chapters in the United States and in Canada. For information about how you may participate in its
programs to preserve wilderness and the quality of life, please address inquiries to Sierra Club, 730 Polk Street,
San Francisco, CA 94109.

First U.S. Edition 1995

First published in Great Britain by Lutterworth Press

Library of Congress Cataloging-in-Publication data is available from Sierra Club Books for Children, 100 Bush
Street, 13th Floor, San Francisco, California 94104.

Produced by Mathew Price Ltd, The Old Glove Factory, Bristol Road, Sherborne, Dorset DT9 4HP, England
Designed by Herman Lelie
Printed in Singapore

10 9 8 7 6 5 4 3 2 1

Who Lives Here?

Maggie Silver

Sierra Club Books for Children

San Francisco

A woodland animal has dug a cozy den
for its family under the roots of a tree.
This quick and clever cousin of the dog
has a reddish-brown coat and a long, bushy tail.
Its babies are called "kits."

Who lives in this safe, warm den?

A group of birds chatters away.
They are weaving grass and twigs into nests
that hang from the highest tree branches.
The male birds are the nest builders.
The females flutter up into secret tunnels inside,
where they lay their eggs.

Who lives in these hanging nests?

In a field, tiny, furry animals
scamper up and down among the wild grasses.
They have woven their nest of straw and leaves
right around the tall grass stems.
These long-tailed, whiskered creatures
are always busy gathering seeds to eat.

Who lives in this ball-shaped nest?

A bird has pecked deep into a tree trunk
to make a hollow for a nest.
The red-headed male did the work.
Then the female laid her eggs inside.
These birds also use their long beaks
to drill into trees and find insects to eat.

Who lives in this hollowed-out tree?

A tiny bird has made a nest
of spider silk and soft grasses.
It can hover next to a flower
and reach in with its long beak
to drink the sweet nectar.
The wings of this jewel-colored bird
beat so quickly that they make
a humming sound.

Who nests in this flowering bush?

Among the leaves of the tropical rain forest,
a family of huge, hairy animals settles down to sleep.
They break up branches to make new nests each night.
These creatures are gentle and intelligent,
and they can stand up on two legs
to reach for a meal of fruit or leaves.

Who sleeps in this leafy forest nest?

High in a tree, hundreds of insects
are busily buzzing.
They are making a special paper
and using it to build their nest.
These busy insects are yellow and black,
but they don't make honey,
and they aren't bees.

Who lives in this papery nest?

A group of furry, brown animals
has built a dam of mud and branches
with a house, called a lodge, on top.
Using their strong front teeth,
they cut down small trees.
They use the wood for food
and for building and repairing their dams.

Who lives in this snug winter lodge?